FIRST TECHNOLOGY
Machines

Author: **John Williams**
Photographer: **Zul Mukhida**

Pistons pumping up and down,
Wheels whirring round and round,
Engines revving loud and brash,
Drills and diggers, hum and crash.
Steam hissing, cogs clicking,
Oil slipping, clocks ticking,
Lights flashing on and off,
I wish all this noise would stop!

Wayland

FIRST TECHNOLOGY

Titles in this series

Energy

Machines

Tools

Wheels and Cogs

WARNING: Machines can be dangerous and must always be handled with care. Young children should always be supervised when using machines. Machines must always be switched off and unplugged after use.

© Copyright 1993 Wayland (Publishers) Ltd

First published in 1993 by
Wayland (Publishers) Ltd
61 Western Road, Hove
East Sussex BN3 1JD, England

British Library Cataloguing in Publication Data

Williams, John
Machines – (First Technology Series)
I. Title II. Series
621.8

ISBN 0 7502 0649 7

Typeset by DJS Fotoset Ltd, Brighton, Sussex.
Printed and bound in Turin, Italy, by Canale.

Series editor: Kathryn Smith
Designer: Loraine Hayes
Photos organized by Zoë Hargreaves

Poem by Catherine Baxter.

Words printed in **bold** appear in the glossary on page 31.

A machine is something that does a **job** for you.

A drill makes holes in wood and metal.

A sewing machine stitches material together.

A hair dryer dries hair.

4

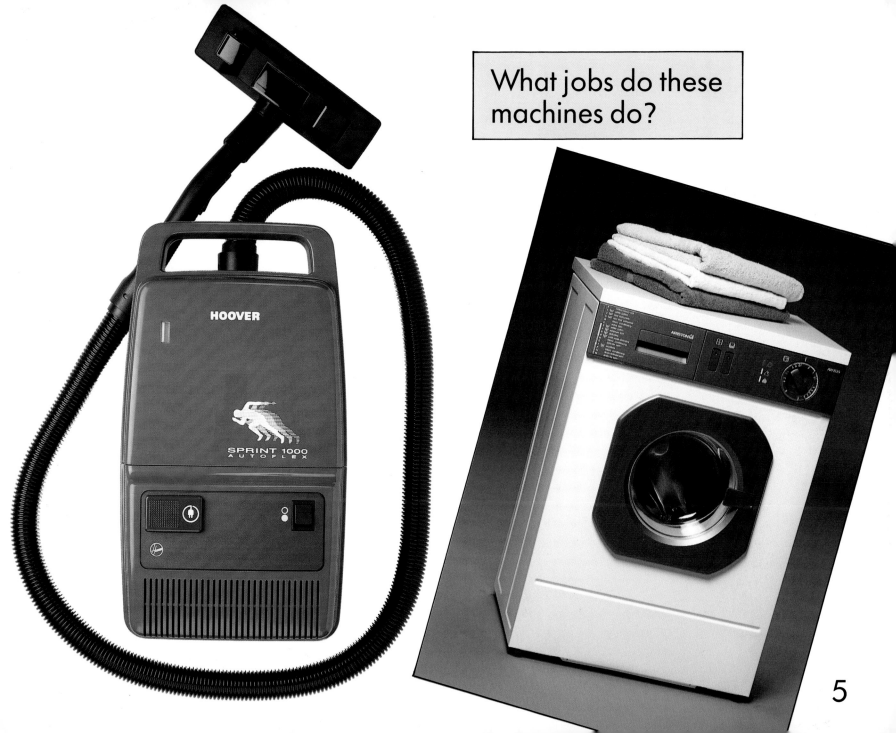

What jobs do these machines do?

HOOVER

SPRINT 1000
AUTOFLEX

5

Machines are everywhere!

Telephones help us at work.

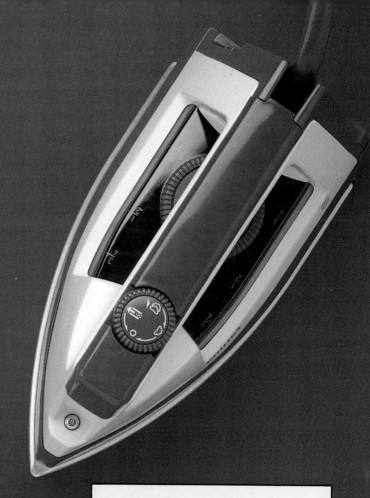

We use an iron
at home to press
clothes.

Machines are useful in the garden.

We even use them to travel from place to place.

Machines help us do jobs that are difficult to do with our hands or a **tool.**

Machines also help us to do things more quickly.
A typewriter helps us to write letters quickly and neatly.

Some machines are very large. This roller coaster looks great fun!

Some machines are very small.
This computer game can fit in your pocket.

Some stereos can be carried in your hands.

Often machines can be dangerous. You should not go near them when they are working, unless a grown-up is with you.

Be careful when using machines that can be easily damaged. Always remember to put them away carefully.

Each machine does a special job.
This machine puts labels on food cartons.

Machines cannot do everything. You cannot use a fridge to cook your dinner. Why not?

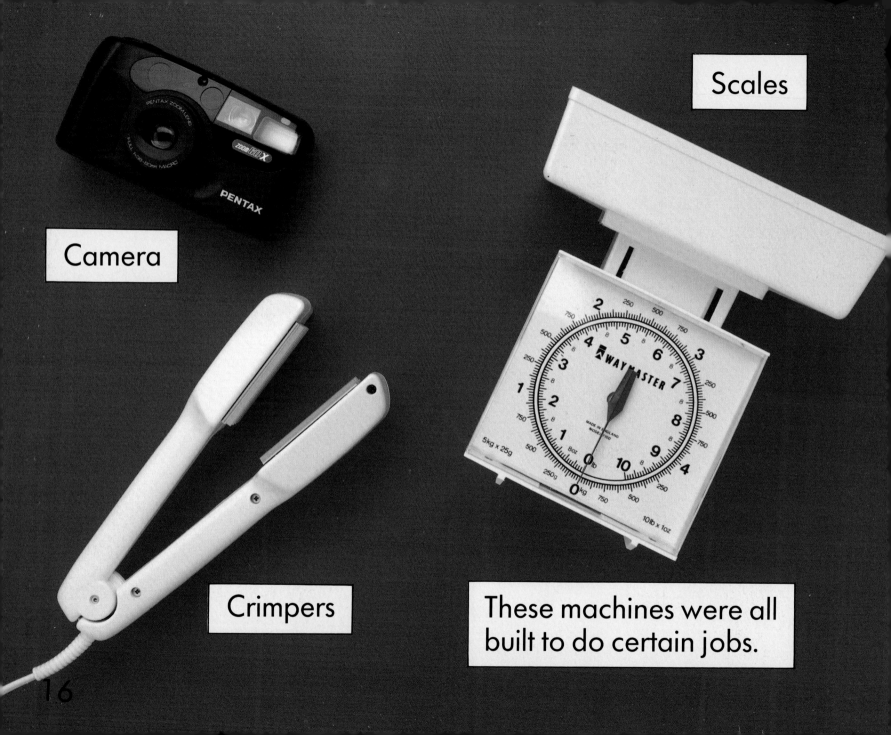

Camera

Scales

Crimpers

These machines were all built to do certain jobs.

16

Which of the machines on the last page would you use to do each of these jobs?

Weighing

Crimping

Photography

Here are some machines
people use at home.
Can you think of any more?

18

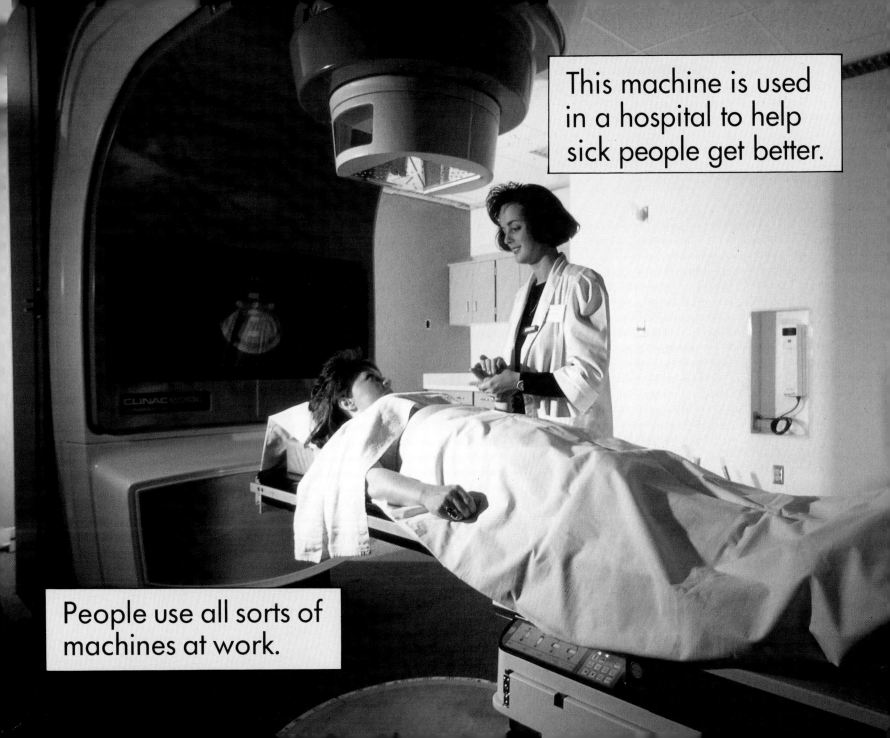

What are these machines used for?

Can you spot any machines you use in your classroom?

What goes on inside a machine? Many machines have **levers** inside.

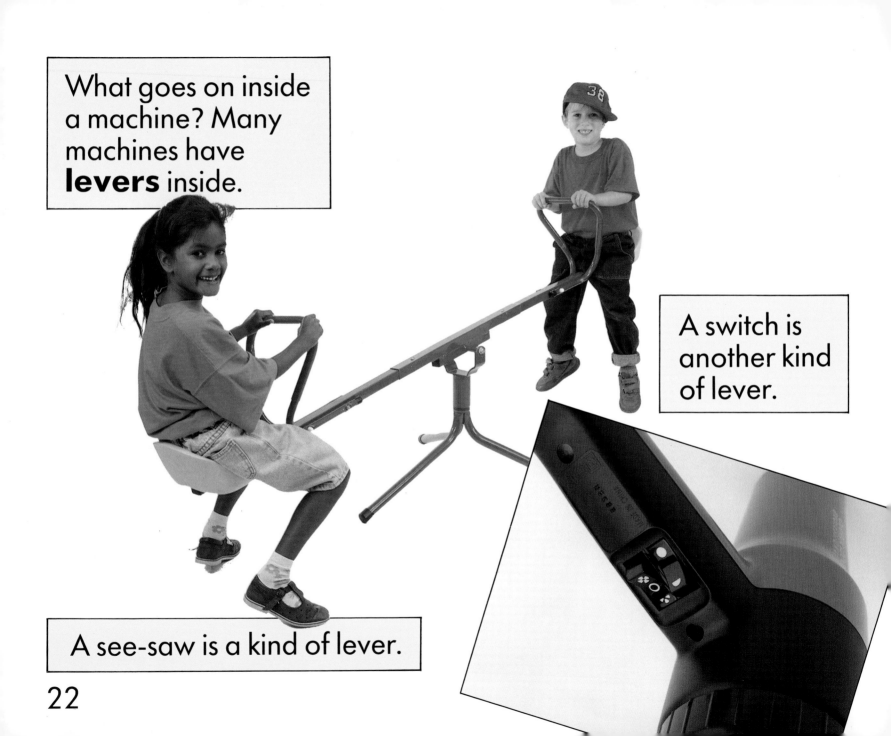

A see-saw is a kind of lever.

A switch is another kind of lever.

Cogs make this
clock work.

The cogs fit together.
When one turns, they all
move to turn the hands
on the clock.

Machines need something to make their insides work.

Cars need **petrol**.

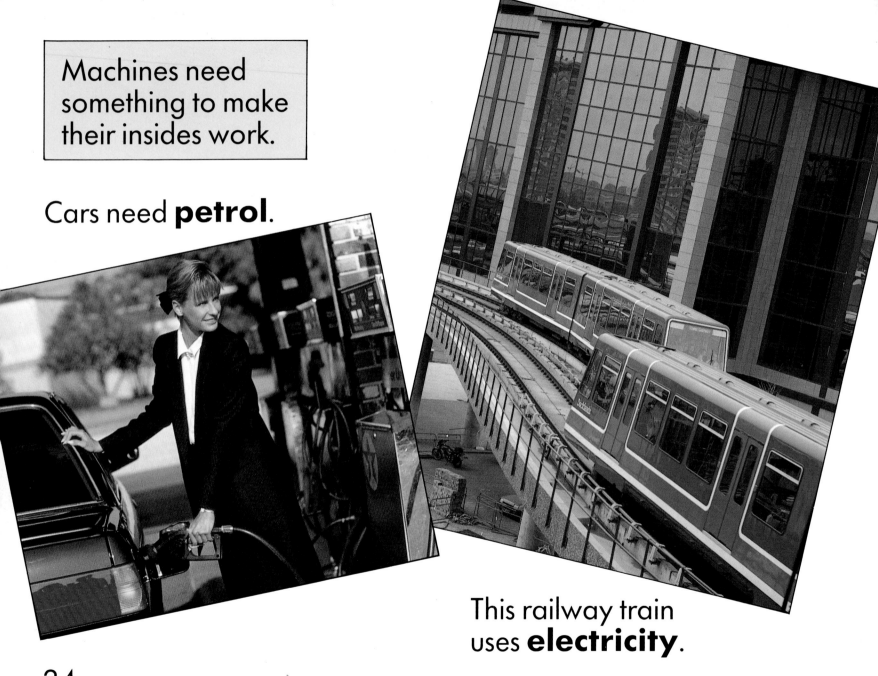

This railway train uses **electricity**.

You only need to push the pedals to make this pedal car work.

This toy works by clockwork.

Original
Kettcar

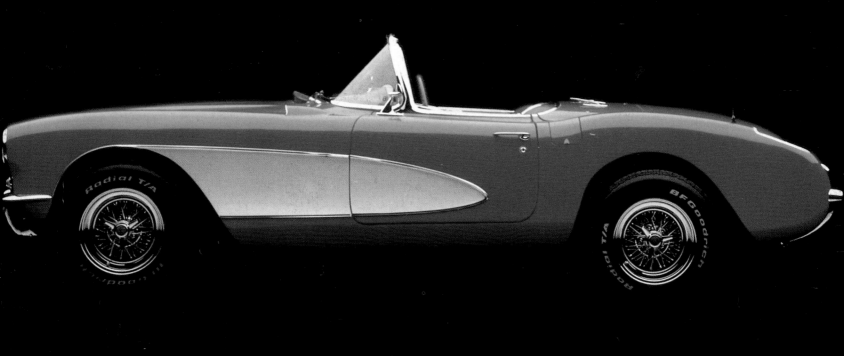

All machines need people to work them.
A car needs someone to drive it.

Other machines only need to be switched on and **plugged in.** Electricity makes this deep fat fryer work.

Building a machine to catch a thief

You will need an old cereal box, a thin strip of stiff card, a marble, a paper clip, a piece of cotton, a tin plate, some sticky tape, a butterfly clip, a blob of sticky tack and a sweet in a wrapper.

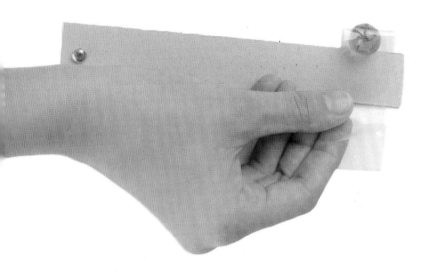

1. Stick the marble to one end of the card strip.

2. Fasten the strip to the top corner of the box with a butterfly clip, so that the arm can swing down easily.

3. Stick the tin plate to the bottom of the box, with one side sticking out.

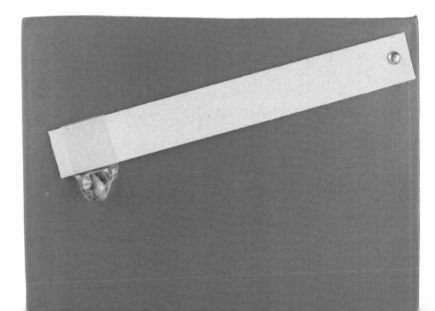

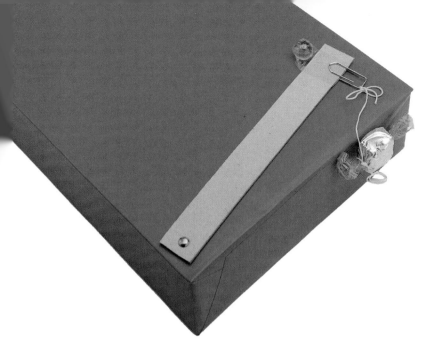

4. Tie one end of the cotton to the sweet wrapper and one end to the paper clip. Slip the paper clip over the card arm, near the marble. Stick the sweet on top of the box with the sticky tack.

5. Can you take the sweet without getting caught?

NOTES FOR TEACHERS AND PARENTS

Children need to learn about machines at an early age to help develop exploratory skills and their understanding of the world around them. They need to be able to identify the difference between machinery and simple hand tools, and recognize various simple machines, what they do and at what cost.

It is also important for children to have an idea of the various forms of energy which drive machines. Although energy is an abstract concept, children of this age should be encouraged to identify machines driven by some of these forms of energy.

Building a machine

The building activity is included to allow children to experiment with design, and to encourage the development of practical skills and their own ideas.

The 'machine' in this experiment is operated by the downward movement of the bar, involving kinetic energy. Although this form of energy is more difficult to understand than the more obvious kinds of energy, such as mechanical, chemical or electrical, children will understand the concept better if, when young, they are given practical work which involves kinetic energy.

GLOSSARY

Cogs Wheels that have teeth round the outside. These fit into other cogwheels so that when one turns they all turn.

Electricity Energy which makes some machines work.

Job Work that has to be done.

Levers Bars used for lifting weights.

Petrol A special liquid used in motor car engines to make them go.

Plugged in When the plug to a machine is pushed into its socket, making the machine work.

Tool Something that is held in the hand, and has been made for a special job.

Unplug To take a plug out of its socket.

Acknowledgements
The publishers would like to thank the following companies for the loan of equipment used in this book; Argos and Seeboard.

SEEBOARD
Doing a power of good

All photographs in this book are by Zul Mukhida, except for the following; Skjold 10; Tony Stone Worldwide 12, 19, 20 (right), 26; Zefa 5 (right), 7 (right), 14, 18, 20 (left), 23 right, 24 (both).

INDEX